Hodder Cambridge
Primary Maths
Foundation Stage

The Stripy Sock

Ann and Paul Broadbent

This book belongs to

....

HODDER
EDUCATION
AN HACHETTE UK COMPANY

Max is searching for a pair of socks to wear.

Find the matching pairs.

Can you point to pairs of socks that are the same?

Sort the socks by colour, size or pattern.

Can you find all the striped socks? Spotted socks? Long socks? Short socks? Green socks?

Max hunts high and low.
He looks in every corner
of his room.

Read the numbers 1 to 5 on the number track.

Can you say the numbers with me? Can you say the numbers on your own

Can you find each of these numbers? 5, 3, 4, 1, 2.

Count the toys.

How many cars? Robots? Bricks? Balls?

How many purple toys? Orange toys? Brown toys?

Max goes outside to look on the washing line.

Find 1 more and 1 less.

How many items are on the line? Add 1 T-shirt. How many items now?

Count all the pegs. Jasmin takes 1 peg. How many pegs are left?

Max is very puzzled.
Where is the sock hiding?

Talk about zero.

Nothing is zero. It is written 0.

What is in the washing basket? Which flower has no petals?

Max has searched all morning. It has made him very hungry.

Compare objects for size and length.
Which glass is big? Which plate is small?
Which spoon is long? Point to a short spoon.

Add objects together to make totals of 5.

How many bananas? How many apples? How many altogether?
How many chairs altogether?

Bel is making sock puppets.

Describe the line patterns.

Follow the lines of the curved patterns with your finger.

Can you find patterns with straight lines?

Max thinks his sock is lost.

Look for solid shapes.

Can you find another object the same shape as the glue-stick?

Can you find objects that are box shaped and ball shaped?

It is now evening
and time for bed.

Talk about the times of the day.

What time of day is this?

What was Max doing at other times in the day?

Describe flat shapes.

Can you find shapes that are squares?

Circles? Rectangles? Triangles?

Count and colour …

1 toy yellow
2 toys green
3 toys purple
4 toys red
5 toys blue

Count the toys above. Write how many.

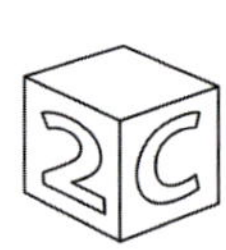

How many T-shirts are on the washing line?

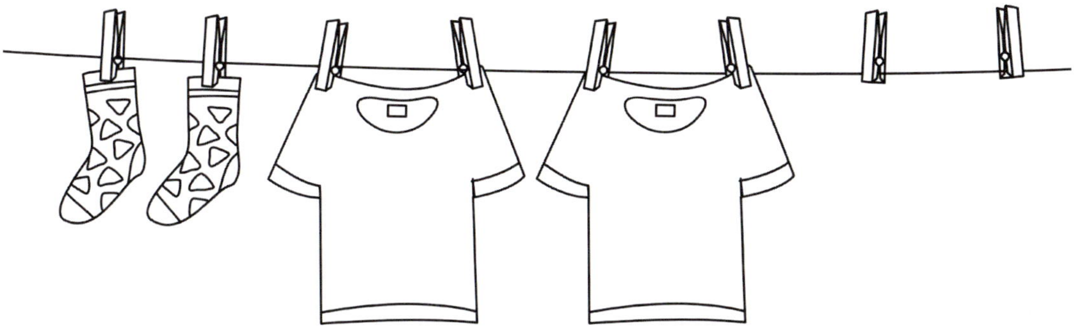

Draw 1 more T-shirt on the line. How many T-shirts are there now?

Colour all the long socks red.

Colour the objects that are box shaped.
Tick the objects that are ball shaped.

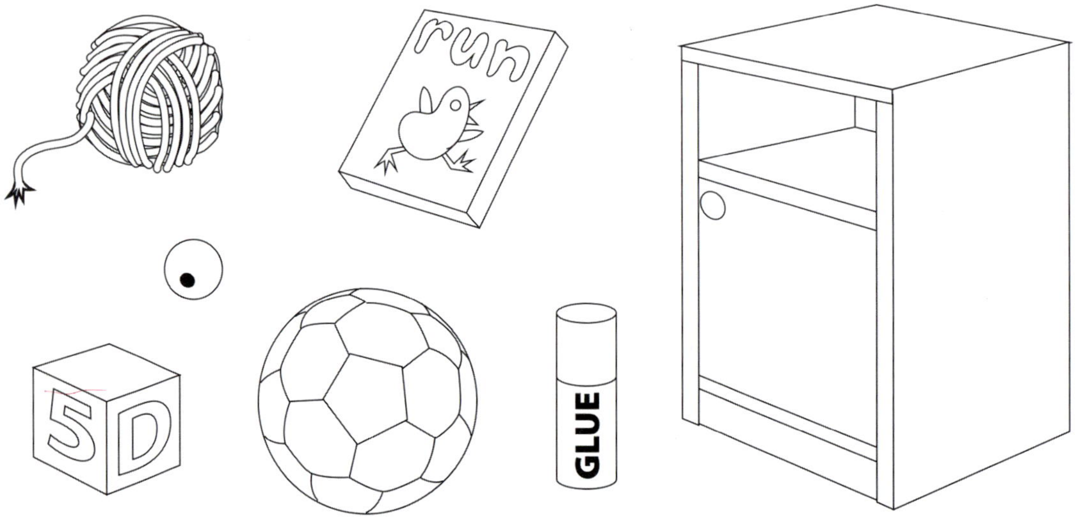

Join these spoons in size order.
Start with the smallest spoon.

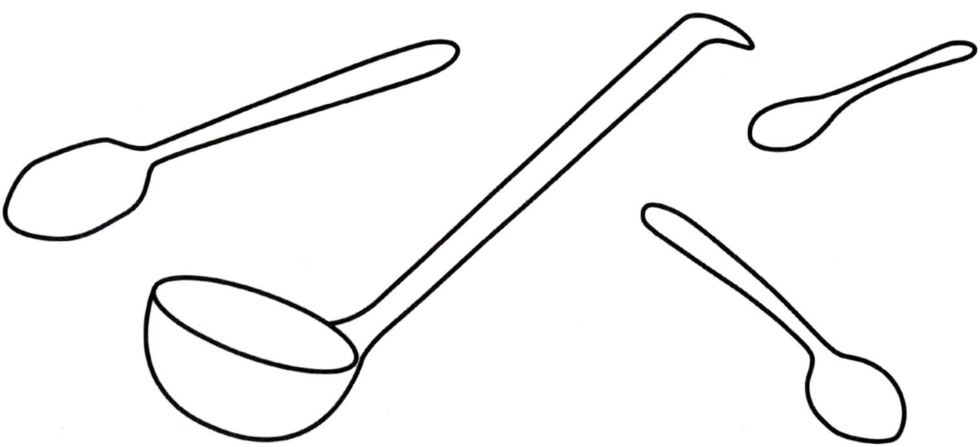